DATE DUE

551.45 BC#0558138
BAR   Barnham, Kay
       Coasts

SPENCER CAMPUS SCHOOL

Coasts

Author: Barnham, Kay
Reading Level: 4.9
Point Value: 0.5
Quiz Number: 78537

Accelerated Reader

Coasts

Kay Barnham

BLACKBIRCH PRESS

THOMSON GALE

San Diego • Detroit • New York • San Francisco • Cleveland • New Haven, Conn. • Waterville, Maine • London • Munich

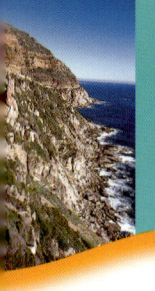

Geography First

Titles in this series
Coasts • Mountains • Islands
Maps and Symbols • Rivers and Streams • Volcanoes

THOMSON GALE

© 2004 by White-Thompson Publishing Ltd.

Produced by White-Thompson Publishing Ltd.
2/3 St. Andrew's Place
Lewes BN7 1UP, U.K.

For more information, contact
The Gale Group, Inc.
27500 Drake Rd.
Farmington Hills, MI 48331-3535
Or you can visit our Internet site at http://www.gale.com

ALL RIGHTS RESERVED.
No part of this work covered by the copyright hereon may be reproduced or used in any form or by any means—graphic, electronic, or mechanical, including photocopying, recording, taping, Web distribution or information storage retrieval systems—without the written permission of the publisher.

Geography consultant: John Lace, School Adviser
Editor: Katie Orchard
Picture research: Glass Onion Pictures
Designer: Chris Halls at Mind's Eye Design Ltd, Lewes
Artist: Peter Bull

Title page: Boats moored off a beach in northern Sardinia.
Page 31: Small fishing boats in Greece.

Originally published by Hodder Wayland,
an imprint of Hodder Children's Books,
a division of Hodder Headline Limited
338 Euston Road, London NW1 3BH

Acknowledgements:
The author and publisher would like to thank the following for their permission to reproduce the following photographs: Corbis 15 (Tim Davis); Ecoscene 9 (John Farmar), 16, 18 (Christine Osborne), 23 (Martin Jones); Hodder Wayland Picture Library *contents page, chapter openers* (Gordon Clements), 11 (Jeremy Horner), 13, 22; Oxford Scientific Films 7 (Scott Winer), 12 (Daniel Valla), 14 (Raymond Blythe), 24 (Steve Littlewood); Photodisc *cover*; Still Pictures 4 (Norbert Wu), 5 (J. J. Alcalay), 17 (R. Leguen), 20 (Andrew Davies), 25 (Cyril Ruoso-Bios), 26 (Claus Andrews), 27 (Truchet-UNEP), 28 (Al Grillo); WTPix *title page*, 19, 31.

LIBRARY OF CONGRESS CATALOGING-IN-PUBLICATION DATA

Barnham, Kay.
   Coasts / by Kay Barnham.
      p. cm. — (Geography first)
Summary: Describes different types of coastal areas, their features, how they are formed, and what lives there.
Includes bibliographical references and index.
   ISBN 1-4103-0112-5 (hardback : alk. paper)
   1. Coasts—Juvenile literature. [1. Coasts.] I. Title. II. Series.

GB453.B37 2004
551.45'7—dc22

2003016793

Printed in China
10 9 8 7 6 5 4 3 2 1

Words in bold **like this** are explained in the glossary on page 30.

Contents

Coasts and change 4
Powerful waves 6
Headlands and bays 8
Beaches 10
Harbors 12
Unusual coasts 14
Coastal wildlife 16
People and coasts 18
Sea defenses 20
New coasts 22
Dangerous coasts 24
Rising sea levels 26
Coast fact file 28

Glossary 30
For more information 31
Index 32

Coasts and change

A coast is a place where land meets sea. Sandy beaches, towering **cliffs**, and sheltered **harbors** are all found on coasts.

The shape of coasts is always changing. Over time, some coasts are worn away. Other coasts are built up by sand and pebbles washed ashore. This can take days, weeks, or even hundreds of years.

▼ *This sheltered bay is at Point Lobos, California.*

Most coasts change shape because of the never-ending movement of the sea. Again and again, the **tide** flows in and out, and **waves** fall onto the shore.

▲ *Jagged rocks jut out along this French coastline.*

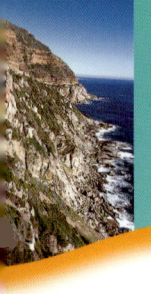

Powerful waves

A wave is a moving ridge of water. Most waves form when wind blows over the sea. The stronger the wind, the bigger the wave. Waves are also created by the tide. This movement of the sea is caused by the pull of the moon and the sun.

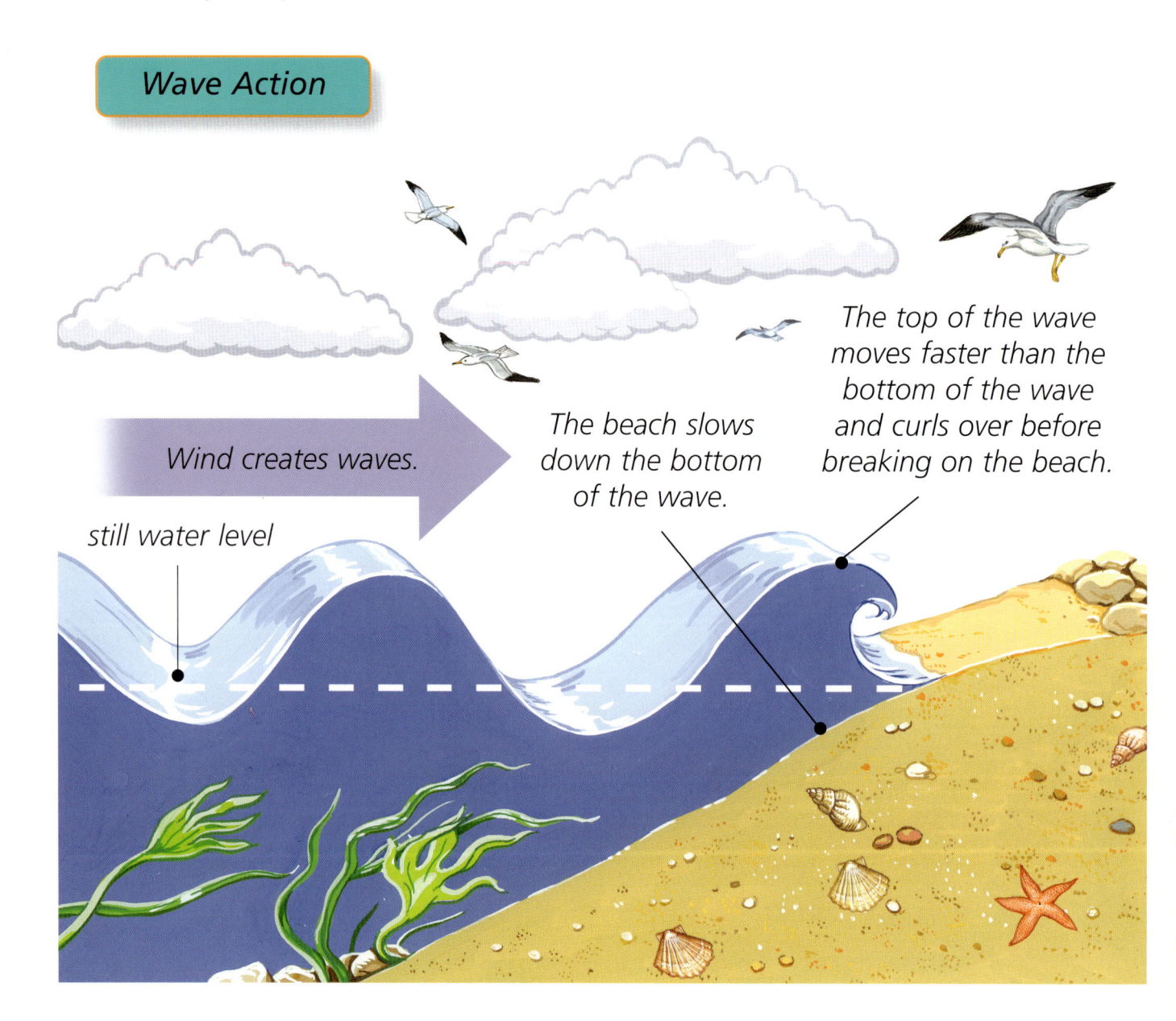

Wave Action

Wind creates waves.

still water level

The beach slows down the bottom of the wave.

The top of the wave moves faster than the bottom of the wave and curls over before breaking on the beach.

Waves can **erode**, or wear away, coasts. The strength of the waves affects how long it takes for this to happen. Crashing **breakers** will erode a coast faster than gentle, lapping waves.

▲ *High waves crash onto the coast of Oahu, Hawaii.*

Headlands and bays

Coasts can be made up of hard or soft rock. Soft rock erodes more quickly than hard rock. Some cliffs are made of areas of hard and soft rock. A bay forms where the softer rock is eroded by waves. Some hard rock may be left sticking out into the sea. This is called a headland.

How Caves, Arches, and Stacks Are Formed

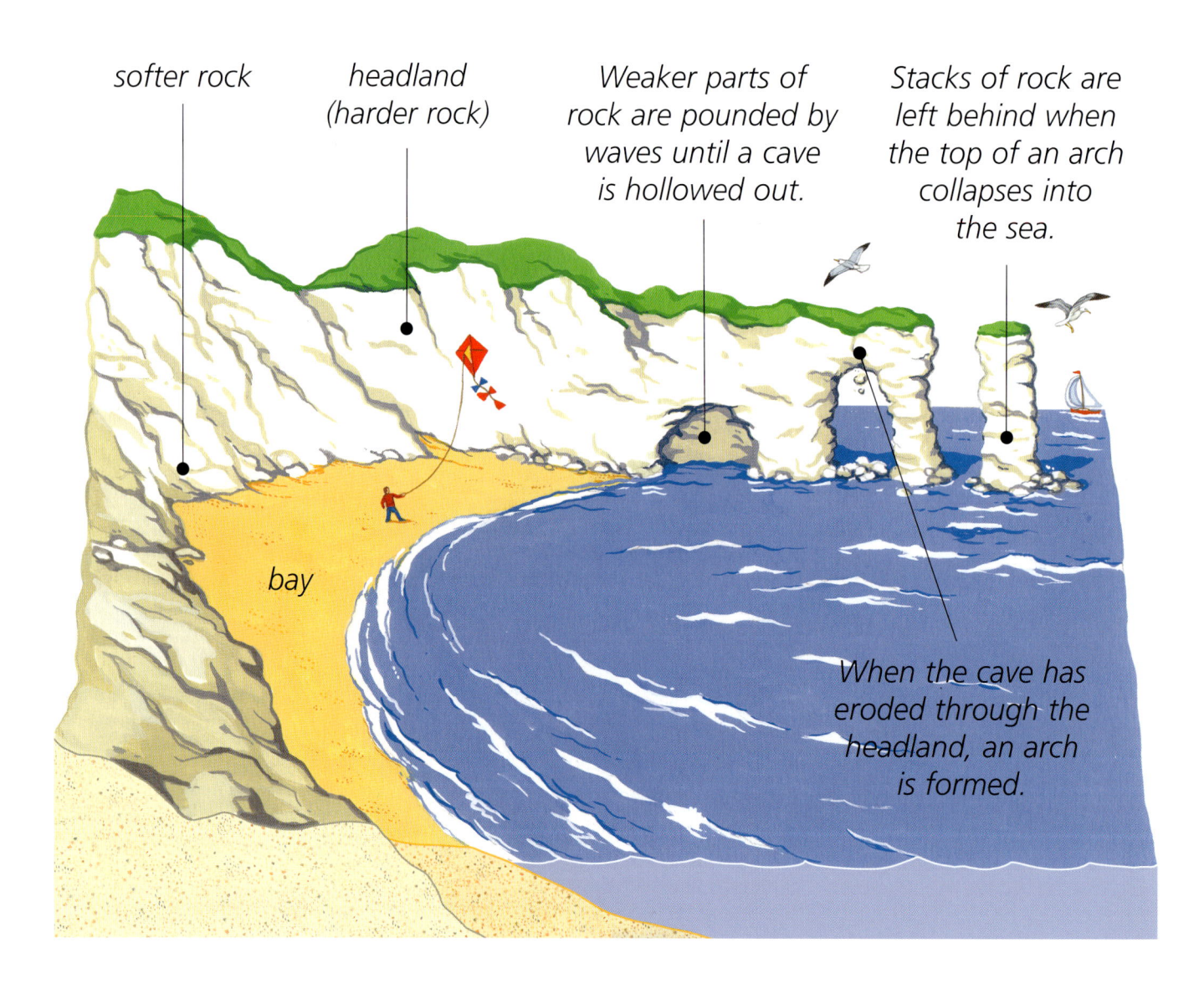

softer rock

headland (harder rock)

Weaker parts of rock are pounded by waves until a cave is hollowed out.

Stacks of rock are left behind when the top of an arch collapses into the sea.

bay

When the cave has eroded through the headland, an arch is formed.

8

Over time, waves wear away the headland itself. Weaker parts of the headland are pounded by the crashing sea. Caves, arches, and stacks are carved out of the rock.

▼ *Dorset, England, has dramatic bays and arches where the sea has eroded headlands.*

Beaches

Beaches are usually made up of pebbles or sand—ground-up rock that has been tossed ashore by the sea. There are two types of waves. Low, gentle waves carry pebbles and sand up the beach, before the seawater slowly drains away. Large, steep, powerful waves crash onto the beach, then drag this material back into the sea.

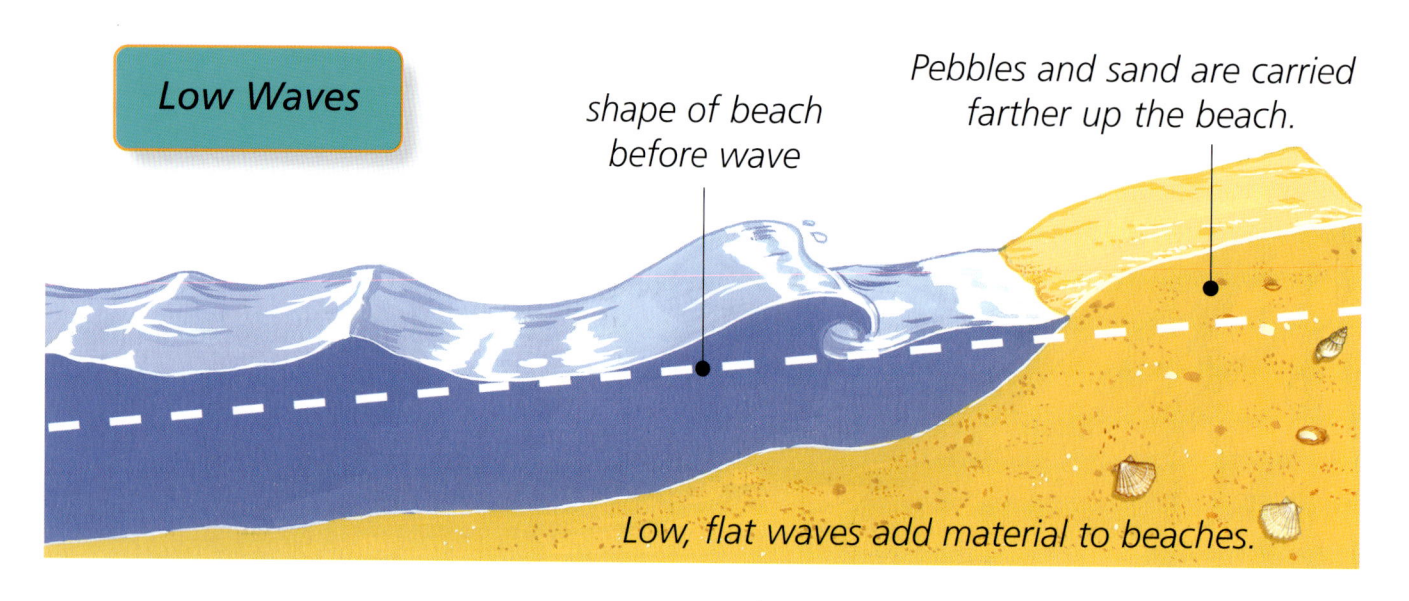

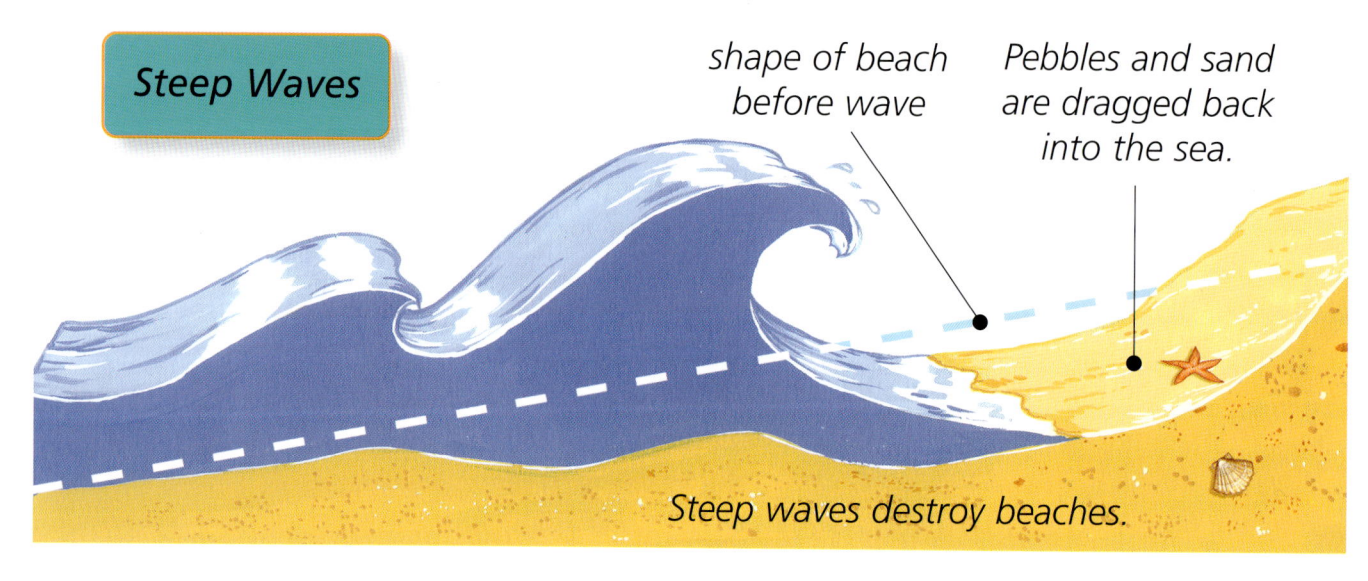

Low Waves

shape of beach before wave

Pebbles and sand are carried farther up the beach.

Low, flat waves add material to beaches.

Steep Waves

shape of beach before wave

Pebbles and sand are dragged back into the sea.

Steep waves destroy beaches.

10

The top of a beach is the highest place the sea reaches at **high tide**. The bottom of the beach is the farthest the sea goes out at **low tide**.

▲ *A wide, sandy, **tropical** beach in the British West Indies.*

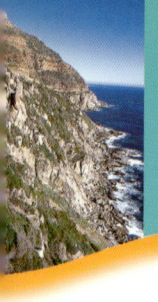

Harbors

Natural harbors are created when waves wear away the coast to form a deep bay. Land on either side shelters the harbor from wind. The water is deep enough to allow boats to sail freely in and out. Big ships, yachts, and fishing boats all use harbors.

▼ *Ships and boats can moor safely in this natural harbor in Majorca, Spain.*

Not all harbors are natural. People build **artificial** harbors, too. Sometimes, a **seawall** is built near the harbor entrance to protect the ships and boats from stormy water.

▲ The artificial harbor in Christchurch, New Zealand, is large enough that oil tankers can moor there.

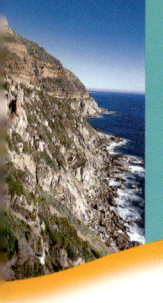

Unusual coasts

Volcanic islands form when undersea **volcanoes** erupt. Beaches on these islands are often made up of very dark sand. This sand is made from the **lava** that erupted from the volcano. It is heavier than the sand that is usually washed onto beaches.

▼ *This beach in Santorini, Greece, is made up of black volcanic sand.*

Antarctica, the continent at the **South Pole**, is completely covered in ice and snow. It is so cold that the sea around Antarctica freezes over. It is difficult to tell where the coast ends and where the sea begins.

▼ *Penguins live next to the icy sea around Antarctica. They dive into the water to catch fish.*

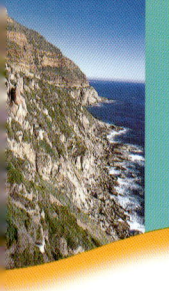

Coastal wildlife

Coasts are home to a wide range of plants and animals. Seaweeds can survive in and out of the water, carried ashore by the tide. Grasses grow a little farther inland. Overhead, seabirds such as gulls and albatrosses swoop and feed, catching fish.

▼ *Puffins nest on a cliff in the Shetland Islands near Scotland.*

Crabs, starfish, and sea snails live in tide pools. Tide pools are holes in the rocks that fill with water when the tide comes in.

Seals, sea otters, and sea lions live in the sea. They clamber onto land to find a mate or to give birth to their young.

▲ *Leatherback turtles live in the sea, but lay their eggs on sandy beaches. When the young turtles hatch, they crawl straight to the sea.*

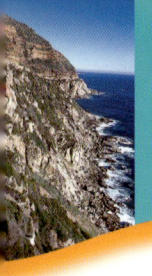

People and coasts

Many people live and work in coastal areas. Some people simply enjoy living beside the sea. Others depend on the sea for work. Fishermen catch fish, which they sell to markets, stores, and restaurants.

▼ *Fishermen check their fishing nets in Zanzibar, Tanzania.*

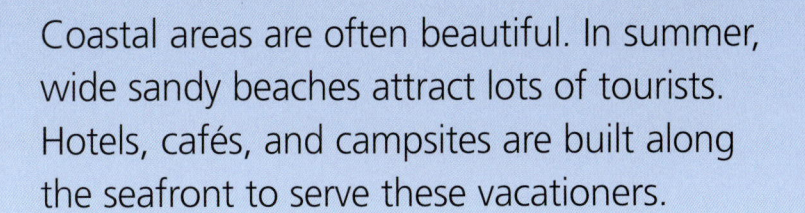

Coastal areas are often beautiful. In summer, wide sandy beaches attract lots of tourists. Hotels, cafés, and campsites are built along the seafront to serve these vacationers.

Building in coastal areas can harm the **environment,** however. Also, if there are too many people around, wildlife can be frightened away.

▼ *Resorts like this one in Australia can be very busy places. Hotels and apartments are built to cope with all the people.*

Sea defenses

Sometimes, the sea erodes coastal settlements and places of natural beauty. When the sea pounds against cliffs, they can crumble. Waves crash against the shore, carrying away sand, earth, and pebbles.

▼ A sandbank prevents the sea from flooding the flat land of Cardiff Bay, Wales.

Rock Defenses

The lighthouse would be in danger without the protection at the foot of the cliff.

Hard rocks are piled up at the base of a cliff.

The sea hits the rocks rather than the bottom of the cliff.

Different kinds of sea defenses are used to protect the coast from erosion. Seawalls prevent the sea from flooding the land. Rocks placed at the foot of a cliff keep waves from smashing against the cliff itself. Sea defenses can cause problems, however, if they keep pushing waves toward other parts of the coast instead.

New coasts

In some places, there is not enough flat land to build new buildings. Earth, stones, and rocks can be pushed into the sea to make new land. This is called **reclaimed** land. Around the world, homes, buildings, and even airports are built on reclaimed land. The new coastlines are usually very straight.

▼ *Shatin New Town in Hong Kong is built on land that has been reclaimed from the sea.*

Some coasts are so low-lying that they are in constant danger of being **flooded**. **Dikes** (wide walls) are sea defenses that are built to stop the sea from flowing over the land. Storm barriers can stop low-lying areas from flooding during rough weather.

▼ *Storm barriers like this one protect the Dutch coast from very rough seas and high tides.*

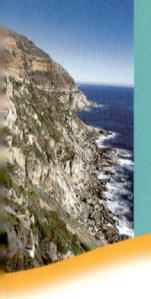

Dangerous coasts

Many coasts have hidden dangers. Jagged rocks under the water's surface can damage and sink a boat. Strong tides and underwater **currents** can quickly drag swimmers into deep water.

▼ *Coast guards train regularly to rescue sailors and swimmers who are in danger.*

Some coasts are well known for their stormy weather and rough seas. At night and in fog, lighthouses shine bright beams of light out to sea, warning sailors that they are near dangerous coasts. Bright plastic floats called buoys mark a safe route through the rocks, guiding boats back to shore.

▲ *Waves batter a lighthouse in Le Havre, France. These waves could force a boat onto dangerous rocks.*

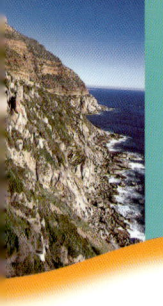

Rising sea levels

The future of the world's coastlines is under threat. Many scientists believe the **atmosphere** (the air around Earth) is slowly getting hotter because of **pollution**. This **global warming** could melt some of the ice at the **North** and **South Poles**, causing sea levels around the world to rise.

▼ *As temperatures become warmer, ice in polar regions melts, and sea levels rise.*

As the sea levels climb, some low-lying coastal areas may be flooded forever. Many countries are trying to slow down global warming by reducing their pollution levels. New sea defenses are planned to protect coasts from the rising sea.

▼ *Beautiful, low-lying islands, like Bora Bora in French Polynesia, are in real danger from rising sea levels.*

Coast fact file

1. Canada is the country with the longest coastline in the world, measuring 151,217 miles (243,360 km), including islands.

2. The country with the shortest coastline in the world is Monaco, at 3.5 miles (5.6 km).

3. The highest sea cliffs in the world are on the Molokai coast in Hawaii. They are 3,313 feet (1,010 m) tall.

4. On March 25, 1989, the *Exxon Valdez* oil tanker created the worst coastal oil spill (below). About 1,491 miles (2,400 km) of Alaskan coast were ruined, and huge numbers of wildlife wiped out.

5. The highest tide in the world takes place in the Bay of Fundy, New Brunswick, Canada. The tidal range (the height difference between high and low tide) can be as much as 52 feet (16 m).

6. The **tidal power station** on the River Rance in Brittany, France, is the only tidal power station in the world. It generates enough power to light up a small town.

7. The largest tidal bore—a wave caused by a tide—was about 29 feet (9 m) high and 1,050 feet (320 km) long. It rushed over Hangzhou Bay, China, on August 18, 1993.

8. The largest natural harbor in the world is Sydney Harbor, Australia.

9. The world's largest artificial harbor was built at Jebel Ali, Dubai, in 1976.

10 Tsunamis are huge waves caused by underwater earthquakes or volcanic eruptions. In 1703, a powerful tsunami hit the coast at Awa, Japan, causing 100,000 deaths.

11 The coastlines of the world measure 994,194 miles (1.6 million km) in total.

Map of Coast Facts

Numbers on this map refer to numbers in the fact file.

Glossary

Artificial Made by people.

Atmosphere The layer of air around Earth.

Breaker A tall, crashing wave.

Cliff A steep, almost vertical, coastline.

Current Strong, underwater movements of the sea.

Dike A wide wall made of earth, with water on one side.

Environment The world around us.

Erode Wear away.

Flood When water covers land.

Global warming When the temperature of the air around Earth rises because of high pollution levels.

Harbor A place for ships to moor.

High tide The highest level that the sea reaches at the coast.

Lava Molten rock that flows or shoots out of a volcano.

Low tide The lowest level that the sea reaches at the coast.

North Pole The point on the globe that is farthest north.

Pollution Damage to air, water, or land caused by harmful materials.

Reclaim To gain something back.

Seawall A wall built to keep the sea from flooding the land.

South Pole The point on the globe that is farthest south.

Stack A tall piece of cliff that remains when the rest is worn away.

Tidal power station A place that turns the power of the tide into energy that can be used by people.

Tide The rise and fall of the sea on the shore, caused by the pull of the moon and the sun.

Tropical Places that are hot all year round.

Volcano A mountain with a gap through which lava escapes from under the Earth's surface.

Wave A ridge of water that moves across the sea.

For more information

Cole, Melissa. *Tide Pools*. San Diego: Blackbirch, 2004.

Furgang, Kathy. *Let's Take a Trip to a Tide Pool*. New York: Rosen, 2003.

Galko, Francine. *Seashore Animals*. New York: Heinemann, 2002.

Greeley, August. *Sludge and Slime: Oil Spills in Our World*. New York: Rosen, 2003.

Jacobs, Marian B. *Why Do the Oceans Have Tides?* New York: Rosen, 2003.

Prevost, John F. *Atlantic Ocean*. Edina, MN: ABDO, 2003

Van Rynbach, Iris. *Safely to Shore: America's Lighthouses*. Watertown, MA: Charlesbridge, 2003.

Woodford, Chris. *Arctic Tundra and Polar Deserts*. Austin, TX: Raintree, 2002.

Index

All the numbers in **bold** refer to photographs and illustrations as well as text.

A
arches **8**, **9**

B
bays **4**, **8**, **9**, 12, **20**
beaches **4**, **6**, **10–11**, **14**, **17**, **19**

C
caves **8**, 9
cliffs 4, **8**, 9, 20, **21**, 28
coast guards **24**

D
dikes 23

E
environment 19
erosion 7, **8**, 20, **21**

F
fishing **18**
flooding **20**, 21, 23, 27

G
global warming **26**, 27

H
harbors 4, **12–13**, 28
headlands **8**, **9**

I
islands 14, **16**, **27**, 28

L
lighthouses **21**, **25**
low-lying coasts 22, 23, **27**

P
pollution 26, 27, **28**

R
reclaimed land **22**
rocky coasts **5**
rock defenses **21**

S
sand **10**, **11**, **14**, **17**, **19**
sea defenses **20–21**, **23**
sea levels **26**, **27**
settlements 20
stacks **8**, 9

T
tides 5, 6, 11, 22, 23, 24, 28
tourism **19**

V
volcanic coasts **14**

W
waves 5, **6**, **7**, 8, 9, **10**, 12, 20, 21, **25**
wildlife **15**, **16–17**, 19, 28
wind **6**, 12

32